CW01150416

Cartoon by a colleague at the Royal Aircraft Establishment, to welcome Hilda back to work.

Adventures in Aeronautical Design
The life of Hilda M. Lyon

Nina Baker

2020

Crampton-Moorhouse Publishing

Copyright © 2020 Nina Baker

All rights reserved. This book or any portion thereof may not be reproduced or used in any manner whatsoever without the express written permission of the publisher except for the use of brief quotations in a book review or scholarly journal.

First Printing: 2020

Crampton-Moorhouse Publishing

Glasgow, Scotland

ISBN: 9798650270584

This book was largely researched during 2019, the centenary year of the founding of the Women's Engineering Society of which Hilda Lyon was a member.

The society supports women in any branch of engineering at any stage of their career.
http://www.wes.org.uk

To archivists everywhere,
because there is always more to know

CONTENTS

ACKNOWLEDGMENTS..2

Hilda's Childhood and Family..3

Early Career..8

Moving on – to the Royal Airship Works...............................15

Across the Atlantic..21

Research in Nazi Germany...29

Back to England...32

To the Royal Aircraft Establishment.....................................35

Second World War Work..38

Death and Legacy..43

References..50

Hilda M. Lyon's Publications, in chronological order..........52

Technical Reports for ARC...54

Adventures in Aeronautical Design and Research..............55

Image credits..62

The Author..63

ACKNOWLEDGMENTS

I would like to thank Alan Brown and Geoff Butler at the Farnborough Air Sciences Trust archives for their unfailing helpfulness in answering my many queries and looking for materials that often did not exist. Also Giles Camplin of the Airship Heritage Trust for his encouragement and positive comments on an early draft.

I must also thank Hilda Lyon's family members in the UK and Australia, who have been very generous in sharing their family's archives, time and knowledge to give us a better understanding of what Hilda was like as a person. In particular the family's generous sharing of materials from the Enid M. Greenwood Archives. The late Enid Greenwood was an eminent local historian, author and genealogist.

Information has come from many, many sources for this book and any mistakes are of course all my fault.

HILDA'S CHILDHOOD AND FAMILY

Hilda Margaret Lyon was born in the small rural town of Market Weighton, in East Yorkshire, on 31st May 1896. She was born into a family who had been local farmers and grocers for generations. Her father was Thomas Green Lyon (1869-1946), a local farmer, landowner, JP and Hull councillor representing Market Weighton as well as being the proprietor of the town's most prosperous grocer's shop [i]. Her mother was Margaret Green (1863-1934), born in Goodmanham, Yorkshire, where her parents were also farmers.

The shop, above which she was born, now a newsagents-convenience store, still stands in the High Street at the heart of Market Weighton, decades after the eponymous wool Market has long gone. From a small grocer's shop Thomas expanded it to become the principal provider of groceries, via delivery carts, bicycles and later vans, to villages all around the area.

She had a younger brother, Harold William (1894-1970), who eventually took over the shop, and three older half-sisters Annie Gertrude, Eva Mary, and Elsie Ida, from her father's first marriage but, out of the whole family, she seems to have been the only one to have traveled and lived out of the area.

Hilda attended primary school in Market Weighton and one of her schoolmates told stories to her own granddaughter of how the teacher had to stop Hilda from answering all the questions put in class. Hilda had her hand up with the answers before the other children had even realised what the question had been about!

The former T. G. Lyon grocery shop where Hilda was born and raised

Brother and sister: Harold and Hilda Lyon

Hilda was then educated at Beverley High School, Beverley being the nearby major town. She was in the first group (in 1908) of only 19 pupils at this newly-founded grammar school for girls, headed by a Miss Rossiter [ii]. The school was strongly academic and offered scripture, history, geography, modern languages, mathematics, natural sciences, drawing, singing and physical education. She would have had access to a proper science laboratory, which was a facility by no means common in girls' schools in that period.

The school today is very aware of its illustrious former pupil and at their first public 'Open Doors day', in 2017, one of the senior girls with ambitions to become an engineer herself, took the persona of Hilda whilst welcoming visitors.

Beverley High School in 2018

Hilda Lyon (2nd left) in gym slip, at Beverley High School

Hilda Lyon then went to Newnham College, Cambridge, one of the early women's colleges, from which she graduated in the Mathematics Tripos [iii] in 1918 – at a time when the university noted the exam results as equivalent to the males-only degree but did not award actual degrees to women. Sadly, she died before the university eventually decided to retrospectively award degrees to its early female graduates.

Newnham College, Cambridge

Hilda in her Newnham College blazer

EARLY CAREER

Hilda graduated during the final months of the First World War and started her first job in July 1918, as an aircraft technical assistant at Siddeley-Deasy Motor Company which was making aircraft in Coventry. She was almost immediately sent to spend six weeks in August and September with the Air Ministry's "Criticism of strength department" [iv], where it is thought that she was sent to take a course in aeroplane stress analysis, to help her convert her theoretical university maths into a practical application.

During the year that she was at Siddeley-Deasy, the company's chief designer, Major F. M. Green (who had been the chief engineer at the Royal Aircraft Factory, which became the Royal Aircraft Establishment, Farnborough), was working on a single-seater biplane which came to be called the Siskin. The structure was a conventional one for the time, of wood and fabric painted with 'dope' (nitrocellulose in those days), but less-conventionally had a greater wingspan for the upper wing than the lower one. Within a couple of years of its first flight – in May 1919 when Hilda left – the RAF was no longer buying wooden planes and all-metal was the new specification.

In 1920 she moved to George Parnall & Co, an aircraft manufacturer at the Coliseum Works in Park Row, Bristol. Hilda, describing her work at this time to colleagues in the Women's Engineering Society in 1944 [v], made the point that she might spend a year in detailed experiments, analysis and design, doing countless complex calculations with her slide rule, only to have the project cancelled on the decision of a higher authority. Her first job there was on the stresses in the bracing wires of a proposed large biplane bomber, but the war ended before it even got off the drawing board.

Siddeley-Deasy Siskin planes

Parnall Puffin

Parnall Possum

Other projects which never got into production were a triplane and the UK's first attempt at an all metal monoplane. In her first two jobs she "… was the Stress Office and the Aerodynamics Department, responsible only to the Chief Designer for my work and with no assistants to check it …" If one of the more experienced design staff queried the sizes she had calculated she would recheck her calculations and then have to find a way to convince him in 'his language' why she was right and so she "… gradually learned to speak the engineering language though doubtless with a strong mathematical accent …" She would later describe herself at this time as a "country bumpkin entering a foreign country where people spoke a different language", but she was intrigued more by the application of her maths than she had been by pure theory at university. Perhaps coming to see herself more as an engineer, she joined the Women's Engineering Society.

During the period 1920-1924 when she was with Parnalls, the company was working on its Puffin, Possum and Pixie plane designs, and the early design of the Plover. Again, Hilda was the entirety of the stress analysis office, so we can assume that she was called on to do work on all the planes which were in the design and test stages during that period.

The Puffin was intended to meet the RAF's 'Specification XXI', for an amphibious fighter-reconnaissance biplane, with the innovation of one central float. Also unusual were its vertical stabilizer and rudder mounted below the fuselage, apparently to maximise the gunner's field of fire. However, the Puffin turned out to have a number of problems: it was tail-heavy and the float initially caused so much spray that the propeller and radiator were damaged in take off tests. Three Puffins were built to try out solutions to the problems but, with no government orders coming in, it never went into production.

In 1922 Hilda Lyon was elected an Associate Fellow of the Royal Aeronautical Society (AFRAeS). Her application form provided most of the clues for the above description of her early career but is also interesting because of the names of those who proposed and supported her application. Harold Bolas, the chief designer at Parnalls from 1917-1929, was the proposer. Hilda Phoebe Hudson was one 'supporter' and her former boss at Siddeley-Deasy, Major Green, was the other. Coincidentally, the two Hildas were both maths graduates from Newnham College Cambridge and for a while they worked together at Parnalls, although Hilda Hudson left in 1921. It would seem that they remained friends as Hudson credited Lyon that year with checking results on a paper of her own.

Hilda Hudson, aeronautical engineer and friend of Hilda Lyon

The Possum was another experimental plane, this time mainly of a plywood construction and a rather boxy appearance. A tri-plane with its engine within the fuselage, the government specification (9/20) labeled these as 'Medium range postal monoplanes', which they clearly were not, but were intended as experimental bomber designs. This may be the 'large bomber' which Hilda remembered working on but which never got into full production. A version of the Possum first flew in 1923 and a second one was sent to the Aeroplane and Armament Experimental Establishment at RAF Martlesham Heath in 1925.

The Pixie, again a design destined to only see three

made, was specially designed to enter the Royal Aero Club's 1923 Lympne Light Aeroplane Trials for light single-seater planes. The idea of the Trials was to promote private flying and the Pixie was designed in two models, one biplane and the other monoplane. The latter design with its wings set into the base of the fuselage, despite being still made from wood and fabric, had a remarkably modern appearance although still with an open cockpit. Pixie I (or 'A') was designed with longer wings and a smaller engine for distance and economy, whilst the Pixie II had shorter wings and a faster engine for speed. Pixie II did well in the speed trials and later racing events.

A Pixie III with two seats was also built but was not particularly successful. Hilda recalled working on propeller designs at this time and ruefully said of her work on this [v] that:

> "I learned something about propeller design and stressing. I remember that we designed two propellers for a light aeroplane competition; one for high speed and the other of a slightly smaller pitch for take off. That was of course, before the days of variable pitch propellers and when the difference between top speed and take off speed was small compared with modern standards. The aeroplane won the speed race with the propeller designed for takeoff. That destroyed my faith in my calculations of blade settings but later I learnt that more expert propeller designers had had similar experiences".

The Parnall Plover was another fabric-covered plane, designed to meet the requirements of the British Air Ministry Specification 6/22. It was a single-seater biplane with a Bristol Jupiter engine, with either wheels or floats (with wheels protruding through the bottom of the floats). The first prototype flew in late 1922 and ultimately 6 Plovers flew operationally with the RAF.

The last plane likely to have been on the drawing boards, before she left Parnalls in 1925, was the Perch. This was in response to Air Ministry 'Specification 5/24', for a two-seat naval training aircraft for the RAF's advanced training and the Fleet Air Arm's deck landing practice and seaplane conversion experience. Yet again, a conventional wood and fabric biplane, Parnalls had been fortunate to be selected for a shortlist of three companies which were to offer prototypes. The Perch did not fly until 1926, after Hilda had left, but in the end none of the shortlisted companies were offered contracts by the government.

Parnall Pixie III

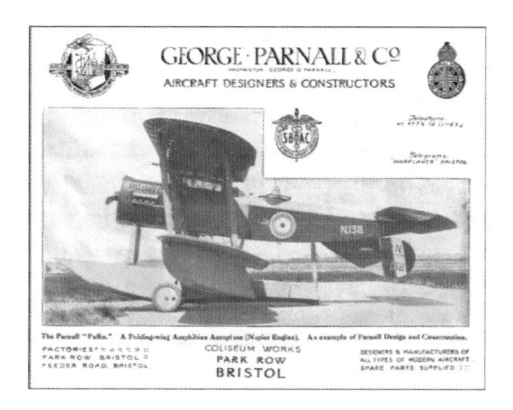

Parnall Perch

Parnall Plover, from a picture in preparation for an unknown publication

MOVING ON – TO THE ROYAL AIRSHIP WORKS

Working on her own in these two relatively small companies had the benefit of early responsibility but it was clear that there was no advancement possible, so she looked around for other opportunities and in 1925, Hilda Lyon joined the Royal Airship Works (RAW) in Cardington as a member of the 'government team' (working on the R101), competing against the 'commercial team' (working on the R100) at Howden, just a few miles from her home town in Yorkshire.

Ironically she never did any work at Howden although the Heritage Centre in the town had a nice little display about the airship works, before the centre had unfortunately to close down. The competition between the RAW at Cardington and the Airship Guarantee Company, a subsidiary of Vickers Ltd, at Howden was to determine whether the state or the commercial sector should design future airships and they were each asked to design a huge airship to the same specification of being capable of carrying troops to India.

Hilda was one of the team of mathematicians and engineers at the RAW working on the design of the R101 airship, her particular role being in stress analysis of the transverse frames. She found the work far more interesting than that required for biplanes and likened the maths to having more in common with the calculations for later aeroplane structures. Airships held a greater fascination for her than the old planes had had and of course there was the added thrill of seeing the framing being constructed in the great shed at Cardington.

The Cardington airship shed

Constructing the framework for the R101 airship

Cardington staff as 'passengers' on the R101

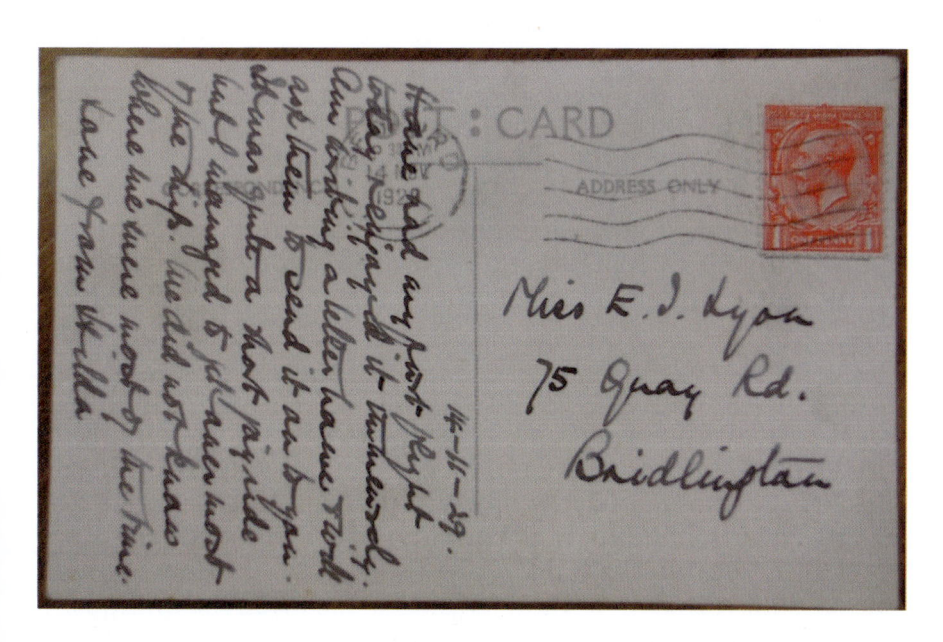

Above and p18, postcard of the R101, which Hilda sent to her half-sister.

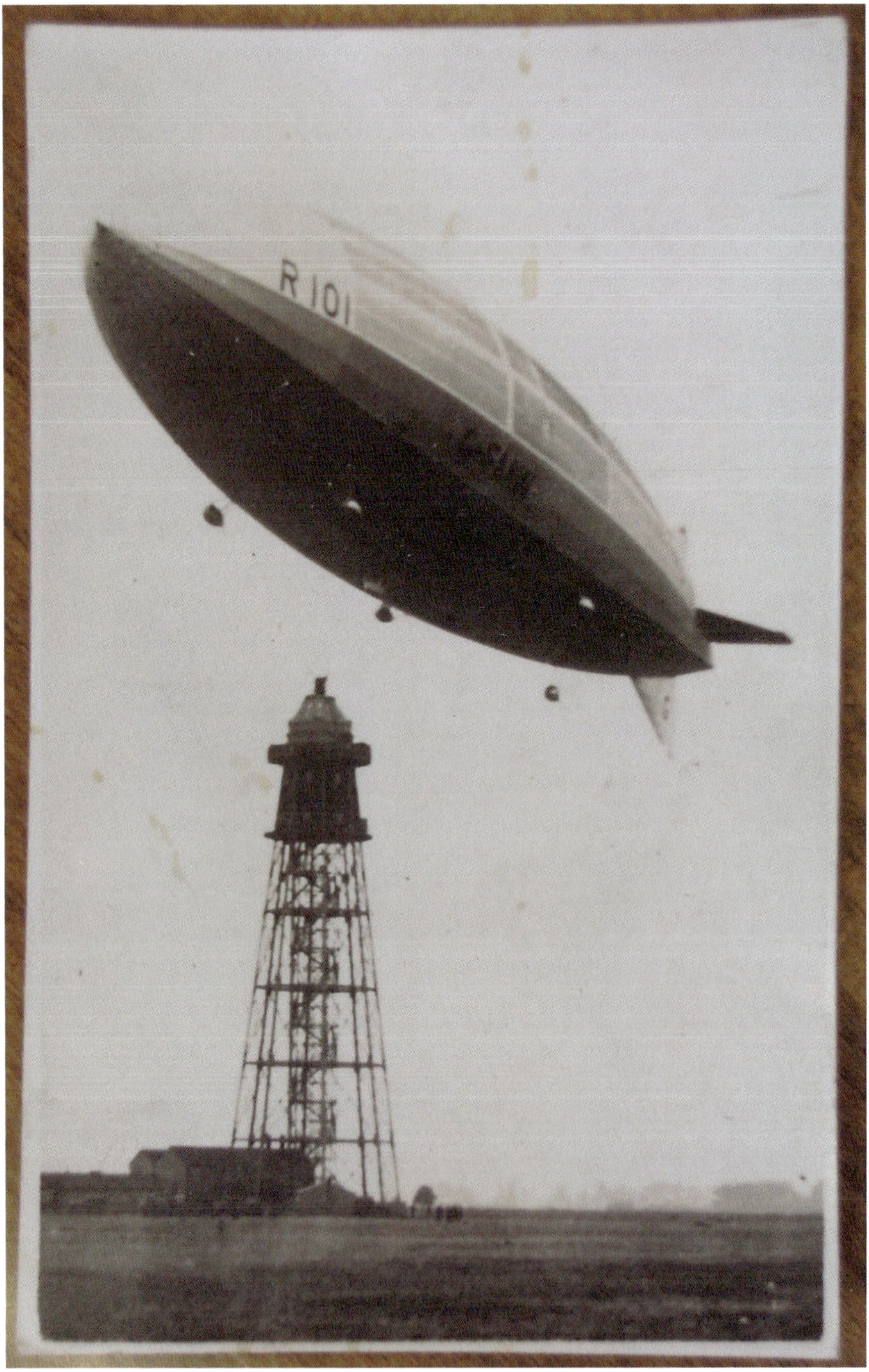
R 101

The 'government team' struggled to design an airship vast enough to contain enough gas to lift the weight of troops called for by the specification, and had to add another section amidships. This addition is not thought to have contributed to its fatal crash at Beauvais, near Paris, on 5th October 1930. Meanwhile, the 'commercial team' at Howden sidestepped that requirement and sent their airship off across the Atlantic, arguably not having met the original specification. Hilda was able to have one 'joy ride' [vi], on the R101's first flight, when, after a long wait for ideal weather; the great airship was towed out of the Cardington shed before sunrise. The airship sat at her mast over the weekend and on the Sunday crowds filled all the surrounding roads for miles around to see her fly for the first time. There are a few photos still in existence taken on the staff 'joy ride' showing men and women in the dining and lounge areas, presumably intended as potential promotional images for the future which was never to be.

Hilda saw the other airship, the R100, return from its successful first flight, on the day she sailed to the USA. The UK's airship programme ended with the crash of the R101 and both airships were broken up for scrap.

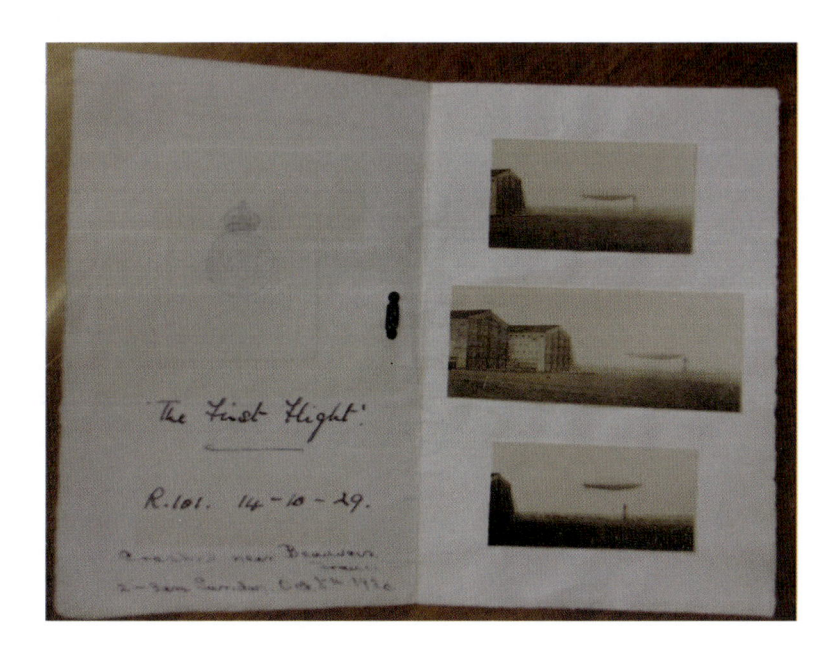

Above and p19, the commemorative booklet which Hilda
Lyon received from the staff flight in the R101.
Note her handwritten comment about the crash.

ACROSS THE ATLANTIC

Although the R101 crash meant the end of the airships programme in the UK, this was not immediately evident perhaps and this period was the start of a fruitful period of research for Hilda, who continued to look at their optimisation. The Aeronautical Journal published her first paper, "The Strength of Transverse Frames of Rigid Airships", in 1930, for which she was the first woman to be awarded the R38 Memorial Prize by the Royal Aeronautical Society. The R38 was the airship that crashed in the Humber Estuary, not all that far from Hilda's home town, on 24 August 1921 with major loss of life. Not far from the cemetery in which Hilda's grave is located, there is a man who has some scraps of the R38's envelope because his father saw it come down when he was a boy and went to scavenge bits.

WOMAN AIRSHIP WORKER.

It was announced at the London and National Society for Women's Service at Westminster last night that Miss Hilda Lyon, a technical officer of the Royal Airship Works, Cardington, had won the Mary Ewart Travelling Scholarship from Newnham College. In August she will go for a year for research work at Massachusetts Institute of Technology.

In 1930 she set off across the Atlantic for two years of postgraduate study at the Massachusetts Institute of Technology (MIT) on a Mary Ewart Traveling Scholarship, awarded by her Cambridge college. She sailed on 13th September from Liverpool, bound for Boston USA on board the White Star Liner Cedric, in a tourist third class cabin.

Her time at MIT gave Hilda her first access to wind tunnels, under the direction of Professor R. H. Smith of the Department of Aeronautical Engineering. In January 1932 she submitted her thesis on "The Effect of Turbulence on the Drag of Airship Models" to obtain her master's degree from MIT.

By the standards of today, the 27 pages of text and another 30 of tables and graphs may seem a bit slim for a master's thesis. However, as a piece of technical research it is surprisingly readable to the interested lay reader. She was a mathematician so, obviously, there is the expected high level of maths involved, but the text explanations of the purpose and applications of her research are probably within the reach of any reasonably focused reader. This may reflect her experiences in Siddeley-Deasy and Parnalls, where she had had to learn how to 'translate' her mathematical results into the 'language' of the engineers. She introduces her choice of research by describing some of the problems that were being encountered at the Royal Aircraft Establishment and National Physical Laboratory:

> "The degree of initial turbulence in the air stream of a wind tunnel has an important effect on the results obtained from measurements of the resistance of airship models. This effect was clearly demonstrated by the international tests on two NPL models, which showed a wide range of values for the resistance of the same model when tested in different tunnels at the same Reynolds number." [vii, p.1]

These differences were revealed to be due to "varying degrees of initial turbulence" and the consequent effects at the boundary layers of the models, with resultant wrong choices as to the optimal shapes, e.g. of airships, based on drag coefficients. Her work was to use models with different shapes but the same fineness ratio (the ratio of length of a body to its maximum width), one based on a shape developed by H. Roxbee Cox (later to become Lord Kings Norton), during the R101 design process and the other model was similar but with a rounder ('bluffer') nose. She aimed to create exceptionally turbulent airflow in the MIT's large (7ft 6in diameter), 60mph wind tunnel, in which the models were suspended from wires. These wires caused her a good deal of difficulty by creating almost as much drag as the models alone. She had to improvise various baffles and duplicates in order to understand the corrections required for the wire drag. The turbulence was created by a series of wire screens in front of the models, the screens' effects being calibrated in front of a sphere, until she found that the screen with the largest mesh had the most effect on the drag of the models.

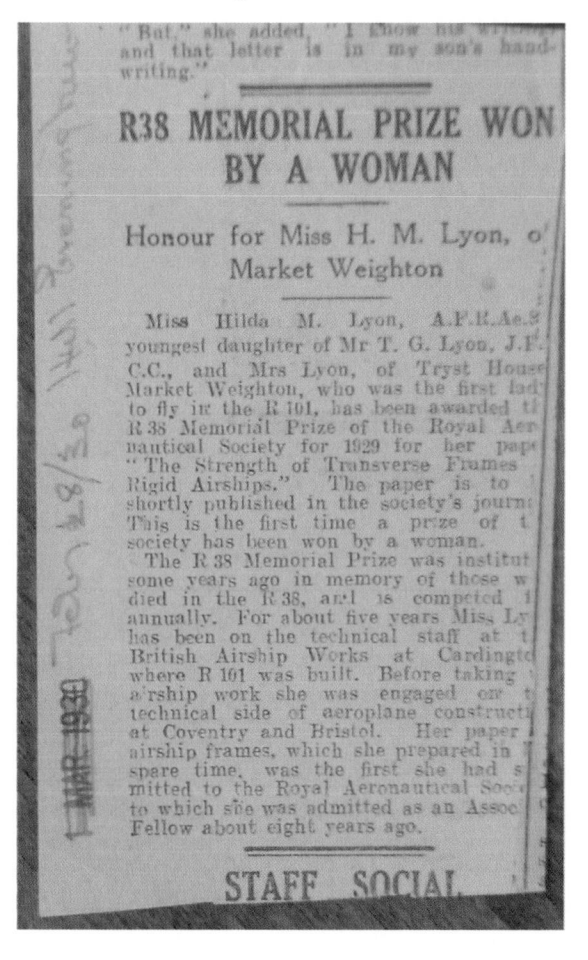

R38 MEMORIAL PRIZE WON BY A WOMAN

Honour for Miss H. M. Lyon, of Market Weighton

Miss Hilda M. Lyon, A.F.R.Ae.S, youngest daughter of Mr T. G. Lyon, J.P., C.C., and Mrs Lyon, of Tryst House, Market Weighton, who was the first lady to fly in the R 101, has been awarded the R 38 Memorial Prize of the Royal Aeronautical Society for 1929 for her paper "The Strength of Transverse Frames Rigid Airships." The paper is to be shortly published in the society's journal. This is the first time a prize of the society has been won by a woman.

The R 38 Memorial Prize was instituted some years ago in memory of those who died in the R 38, and is competed for annually. For about five years Miss Lyon has been on the technical staff at the British Airship Works at Cardington where R 101 was built. Before taking airship work she was engaged on the technical side of aeroplane construction at Coventry and Bristol. Her paper on airship frames, which she prepared in her spare time, was the first she had submitted to the Royal Aeronautical Society, to which she was admitted as an Associate Fellow about eight years ago.

News cutting from 1930 about Hilda winning the R38 prize

THE EFFECT OF TURBULENCE ON THE DRAG OF AIRSHIP MODELS

by

Hilda M. Lyon,

M.A. of the University of Cambridge, England.

Submitted in partial fulfillment of the requirements

for the Degree of

Master of Science

from the

Massachusetts Institute of Technology

1932

Department of Aeronautical Engineering.

January 20. 1932.

Professor in charge of research

Chairman of Departmental
Committee on Graduate Students

Head of Course.

Title page of her thesis

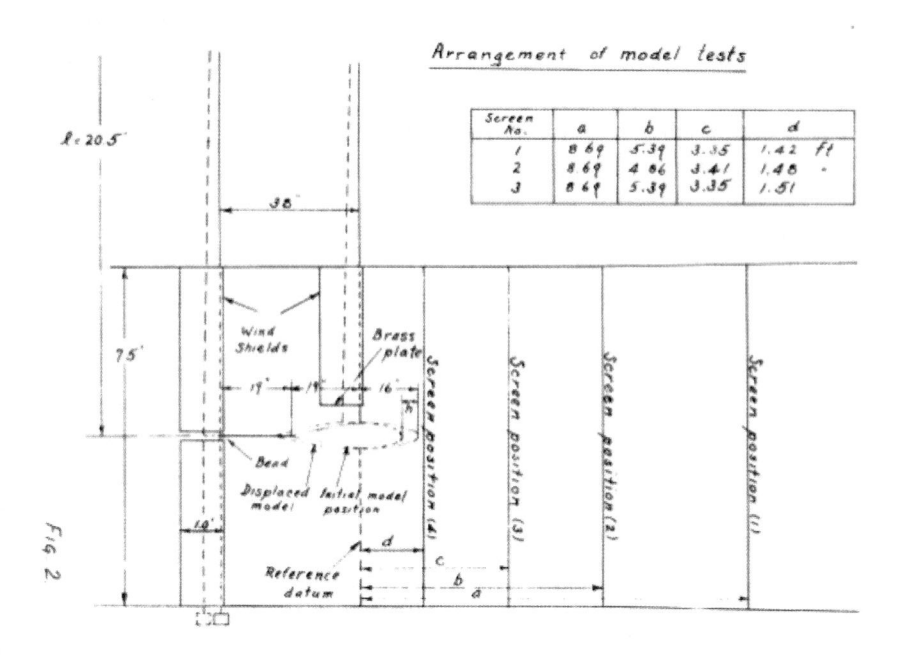

Fig 2 from her thesis, showing how Hilda set up the airship models in the wind tunnel at MIT.

Whilst her results' analysis is necessarily mathematical, her Part VI Practical Application of the Results is a straightforward piece to read even without specialist knowledge:

> "The relative merits of airship shapes are usually estimated in terms of the drag coefficient kd... ...and the ratio is approximately constant for all streamline bodies of the same fineness ratio, and varies as the cube root of the fineness ratio... ... The results shown in Fig. 21 suggest that for the full scale Reynolds number kf has the same value for both shapes. With this assumption two ships of the same volume and the same fineness ratio will have the same drag. In considering the speed and required horsepower there is therefore nothing to choose between the two shapes. From the structural point of view the shape with the higher block coefficient* has a definite advantage, as it provides a greater gas volume in the nose and tail to balance the concentrated weights of the mooring equipment and the fins, thus relieving the static bending moments on the hull and reducing the structure weight. As the structure weight in modern rigid airships is about 65 per cent of the total lift, a saving of 10 per cent on the structure weight means an increase of nearly 20 per cent in the useful load and probably 65 per cent in the 'pay load'." [vii, p.23-24]

---

* The block coefficient of a shape is the ratio of its volume to the volume of a rectangular block of the same overall dimensions. The nearer it is to 1, the closer the shape is to a rectangular block.

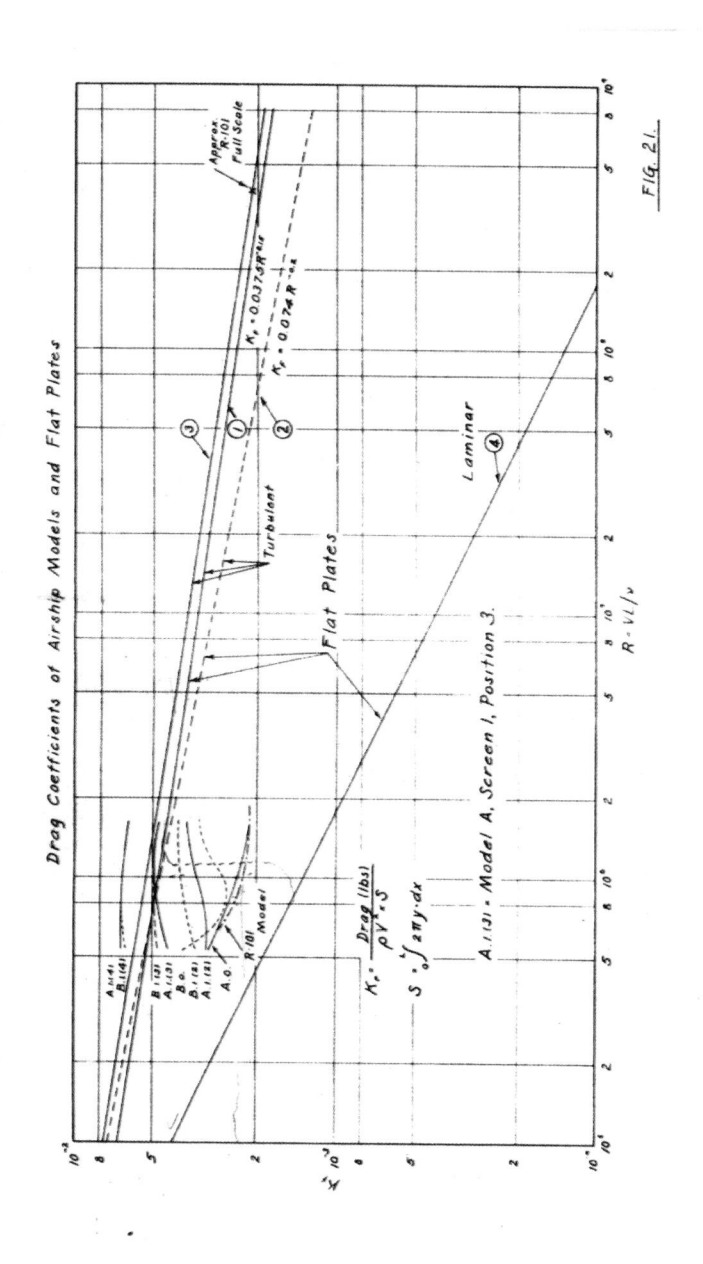

Results graph from her MIT masters' thesis, comparing the performance of various shapes

Of all Hilda Lyon's work and publications, her thesis [vii] is the one most cited by other authors: there have been 20 citations from 1934 right up to 2014. She later said that one of the great benefits she gained at MIT was having to do so much of the practical work herself and that if she had had her career to start again she would have done a practical engineering course at a technical college after gaining her maths degree.

RESEARCH IN NAZI GERMANY

That work got her another Mary Ewart grant which enabled her to go for a year to Göttingen in Germany, to do aerodynamics research at the Kaiser Wilhelm Gesellschaft für Strömungsforschung with Professor Ludwig Prandtl. His research at the time was in the perfection of airflows in wind-tunnels and aerodynamics generally, which would have been the reason she went there. The year she arrived he published several papers [viii] including a chapter on how to attain a steady air stream in wind tunnels, in a textbook on experimental physics, and the following year several more including one on turbulence in wind-tunnels and one on dynamic soaring. Although she said she was largely doing theoretical work, it is possible to imagine that Hilda contributed to some of Prandtl's work, in her role as a junior research assistant.

Amongst the innovations which Hilda would have access to at Göttingen would have been the best wind tunnels in the world, as designed by Prandtl. These were the first to be in the form of a continuous loop rather than a straight tunnel and would be the design for all wind tunnels of the future.

Professor Ludwig Prandtl with one of his water-tunnels in the 1930s

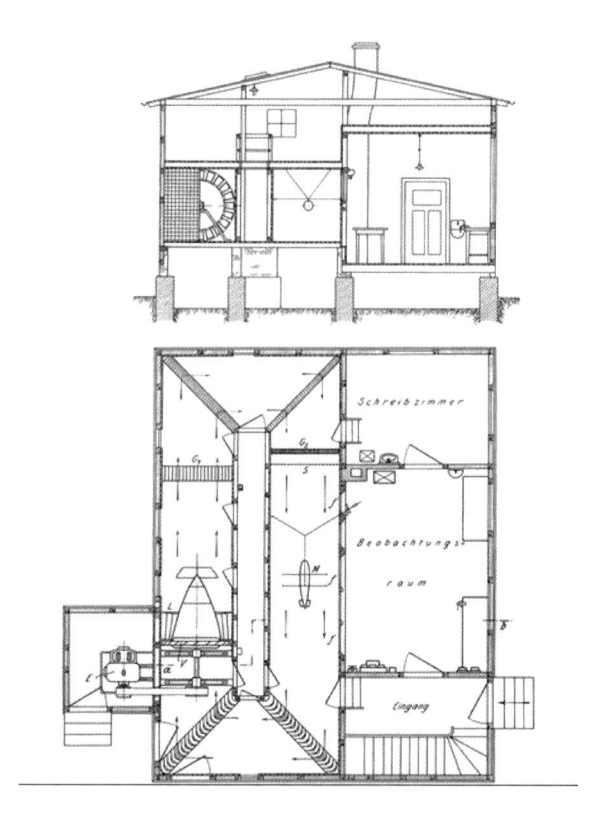

Prandtl's plans for his wind tunnel facility at Göttingen

Hilda was in Germany as the Nazis were ascending to power and the summer of 1933 must have been terrifying as Jewish staff and students were marched out of the university. Most of the students were Nazi supporters and she witnessed one of the infamous Fackelzug torchlight processions. However, although she was fortunate (honoured, even) to be able to go there to work with Prandtl as the foremost expert in her field, he was a Nazi supporter and so safe from the Party's thugs. Interestingly, given this alignment with the Nazis, Prandtl seems not to have shared their views that a woman's place was in the home (Kinder, Kirche, Kuche – children, church and kitchen) and was an active supporter of women in science. His department had a number of women postgraduate students and research assistants, some of whom went on to be important in their own right later on. Unlike many of his colleagues, he was not recruited to the RAE after the war, due probably to age and ill-health and remained at the Göttingen institute until his death in 1953.

Uniformed students marching in Göttingen in 1933

BACK TO ENGLAND

Hilda's time in Germany was cut short when her mother was taken ill and in July 1933 Hilda had to return home to care for her in Yorkshire. Fortunately, this potentially career-killing domestic obligation was partly overcome by her former colleagues at the Royal Aircraft Establishment and others in the field who, understanding her talents, helped her to find work she could do from home, sometimes even being able to pay her for small pieces of work. She was able to use the university libraries in Hull and Leeds and pay occasional visits to the National Physical Laboratory (NPL) in Teddington (west of London) and the RAE in Hampshire, but she really missed the collegiality of being in the midst of colleagues.

Hilda Lyon with her mother

One such piece of work, although it is not known whether she was paid for it, was in 1936 for the founder of the NPL, the physicist Sir Richard Glazebrook, by this time at Imperial College. Their paper "On the force between two coaxial single layer helices carrying current" was published in summary form in the Proceedings of the Royal Society A (Maths, physics and engineering sciences). The great man very generously credited Hilda on the front page. This seems to have been Hilda's only venture out of aeronautical engineering, but clearly her mathematical skills were what Sir Richard valued:

> On the Force between Two Coaxial Single Layer Helices Carrying Current
>
> By Sir RICHARD GLAZEBROOK,* F.R.S., and H. M. LYON, M.A.
>
> (Received September 25, 1935)
>
> The following is an abstract of a paper bearing the above title which is being printed in the Collected Researches of the National Physical Laboratory, vol. 24.
>
> * My share amounts to little more than the evaluation of the integral expression for the force and suggestions as to the method of treatment; the integration is due to Miss Lyon.—R. T. G.

In 1934 Hilda was able to collaborate with William Jolly Duncan, a lecturer in aeronautics at University College Hull in his work on the flutter of wings and elastic blades. Hilda published two papers on streamlining and boundary layer effects by 1934 and two more in 1935, plus a paper on "Oscillations of elastic blades and wings in an airstream" with William Duncan and Arthur Collar in 1936.

Despite not being in formal paid employment in the field from 1932-37, she was recognised in a 1934 issue of Flying Magazine as being "...the classic authority on the subject of stresses in transverse frames".

38,321,196 passenger miles flown per fatal accident. This is a record that the railroads should envy.

THE TEXAS COMPANY has closed a year's contract with the Transcontinental and Western Air for supplying that airline with gasoline and lubricating oil.

CLARENCE L. CLABAUGH has resigned as vice-president of North Shore Airways, Curtiss-Reynolds Airport, Glenview, Ill. J. G. Boess was elected to fill the vacancy.

RADIO COURSE INDICATORS, recently developed by the Westinghouse Electric and Manufacturing Company, remove the necessity for the pilot listening to the radio beacon tone signals. The indicator is similar to an auto dash ammeter so that the signals are read visually.

THE ZMC ALL-METAL AIRSHIP has shown such excellent performance that the Navy contemplates the construction of a similar all-metal airship built to the size of the Los Angeles. All that the Navy lacks for this purpose is money, but it is hoped that a sufficient appropriation will be voted by Congress.

MAJOR ERNST UDET, the stunt pilot, has purchased a Curtiss Hawk single-place pursuit biplane for his acrobatic work. It is powered by a 700 h.p. Wright Cyclone engine and develops a top speed of 206 m.p.h. with a diving speed above 360 m.p.h.

MEXICAN PILOTS flew six Kinner powered Fleet planes from Buffalo, N. Y. to Mexico City. They will be used as training planes for the aviation branch of the Mexican Army.

THE BOEING SCHOOL OF AERONAUTICS, Oakland, Calif., has issued a 36-page booklet that discusses the matter of aeronautical occupations in an interesting manner. It lists all of the known jobs offered by aviation, the duties and requirements.

AIR RACES are apparently no longer an attraction, for little money was made among the major meets. The National Charity Air Pageant, held at Roosevelt Field went into bankruptcy with a deficit of $40,368.

A KANSAS DeSOTO-PLYMOUTH DEALER has a private plane with which he keeps in touch with the DeSoto factory in Detroit, Mich., or with this ship he can talk a prospect into a DeSoto sale anywhere within a 75 mile radius from his store.

END.

Airships
(Continued from page 30)

stresses, the longitudinals are progressively adjusted and recalculated until correct.

For some time, the diagonals caused difficulties. To omit them entirely was clearly incorrect, and to use shear methods instead of bending moments also led to inaccuracies. Professor Wm. Hovgaard, American, found a solution, whereby the diagonals are made a supplement to the longitudinals and the two elements calculated together. His method has been adopted both in the U. S. A. and Great Britain.

For the transverse frames according to the methods used, the aerodynamic forces are of minor importance. Here one deals with dead loads of structure, machinery, equipment, fuel, passengers, and crew, also with the supporting forces of the gas. Even so, the problem is not simple because this simple frame is also indeterminate; however, the elements are fewer, the redundancy not excessive, and aid is available also from the symmetry of conditions.

The classic authority on the subject is Hilda M. Lyon of the Cardington staff, who has written a thorough and competent paper on the stresses in transverse frames—likewise a R38 Award paper. This author admits freely that her solution has reference to the limited conditions assumed only and that there are forces originating in the longitudinals, or from aerodynamic forces passing through them into the transverse frames, but she has no means to evaluate these forces and therefore no means to figure their effects on the frames.

The practical designers have thus placed the three sets of elements into two groups, figuring the direct effects on each group separately, but having no definite or reasonably accurate method to find the interaction between the two groups. The omission is obviously a very serious matter, and to prove this it is sufficient to refer to the reader's own common sense.

Suppose he were to have a small model of an airship frame, were to hold this between his hands, and bend it over his knees. His hands and his knees thus represent aerodynamic forces. One cannot escape the conviction that the bending action will compress the frames in a transverse direction, will flatten them out into an elliptic shape. This clearly means supplementary forces on the transverse frames, and by proportion, very heavy ones.

One may visualise an airship suddenly attacked by violent wind squall and the bending caused thereby—most assuredly the secondary dynamic effect on the transverse frames are going to be more severe than the primary one from loads only. And yet one has made no definite provision for these effects for the primitive reason that one does not know how to figure them. For this reason, possibly, that so many good men have been sent to their deaths?

In the two design papers referred to —one of American and the other of British origin—the necessity of separate treatment has been admitted. It is also implied in the Lyon paper. We should then expect that the authorities check up on a theory, which is admittedly incomplete. We should expect the designers to ascertain the actual stresses in the transverse frames of completed ships during conditions of actual navigation.

On the German built ZR3, now the Los Angeles, extensive tests were made in 1920—see pages 324 and 325 of the United States Air Service. C. P. Burgess, editor of the material pertaining to forces, stresses, and strains. During these tests no less than 35 strain gages were placed to find out in what manner the ship was affected by aerodynamic forces and actual stresses compared with those calculated.

So completely absorbed were the engineers of these tests, in the relation between wind forces and the stresses in the longitudinals, that they applied all their instruments to these—and to these only—if the report as published is really complete. There was no check on the transverse frames, and to this day one does not know what increment to allow in these for the forces coming through from the longitudinals. Were these forces considered in the Akron?

END.

TO THE ROYAL AIRCRAFT ESTABLISHMENT

In 1937 the Royal Aircraft Establishment (RAE) was finally able to create a post for her to work as a full time Principal Scientific Officer (civil servant) in the Aerodynamics Department, initially working in the wind tunnels on boundary layer suction. She then joined the Stability Section, later becoming head of this section, and also served on the Aeronautical Research Council.

From 1937 onwards she was publishing frequently, up to and even after her death (see full list of her publications at the end). Her friend and colleague at the RAE, Frances Bradfield, commented after her death that 'this was the critical period of her career, as a four year break would have ended most women's work: instead of which she came to the Royal Aircraft Establishment, when she was again free, with added range' (xx).

It is worth remarking that the RAE was notable for employing many high-calibre women in its civil service research staff. Its very first female employees, back in its 19th century ancestral Royal Balloon Factory days and up to the 1920s, were artisanal experts – the women of the Weinling family who introduced their technique of making the airship gas envelopes from goldbeaters' skin (treated cow gut). The loss of so many male mathematicians and scientists during The Great War, just at the time when aeronautics was developing so quickly, meant that the establishment started seeking out talented female mathematicians and scientists even before that war ended. Throughout the 1920s and 1930s, when so many hundreds of experienced women engineers were unable to find work during the difficult economic conditions in industry, the RAE pro-actively sought the best mathematical and engineering graduates whether male or female. Not so famous perhaps as the women who were the secret code-breakers of Bletchley Park in the Second World War, but the women at the RAE throughout its history contributed vastly to aerospace development and air safety in the UK and globally.

Pilot Joy Ferguson, RAE engineer Chrystelle Fougere,
Jane Whittle and Miss Brotherton with a Gloster jet at the
RAE

Hilda was by no means a solitary female professional at the establishment.

Some other notable women at the RAE during Hilda's time there included:

Frances Bradfield – wind-tunnel expert, specialising in aerodynamic stability. Bradfield was also notable for being the woman who welcomed generations of new scientists male and female and taught them the basics of how they were expected to work at the RAE.

Anne Pellew (Mrs. Burns) – specialist in aircraft stresses and clear air turbulence, also part of the team that solved the problem of the Comet airliner crashes.

Chrystelle Fougere (Mrs. Somerville) – aerodynamicist specialising in performance of jet aeroplanes and early supersonic planes.

Helen Grimshaw – human engineering specialist, helmets, pressure suits (and early ideas for space suits), de-icing equipment.

Norah Searle (Mrs. Irving) – aerodynamicist.

Kate Maslen - very early development of wire strain gauges in testing large aircraft components. Also breathing systems and helmets for high altitude fast jet fighters (and early space capsules), and effects of noise and vibrations on human performance.

Beatrice Shilling (Mrs. Naylor) – famous for her work correcting fatal problems in the Rolls Royce Merlin engines powering the Spitfires, but whose work on post-war rocketry was arguably more important.

SECOND WORLD WAR WORK

Hilda Lyon's work in the Second World War included stability analysis of the Hawker Typhoon rudder and elevator. This was also the period when she did some of her most enduring work: on longitudinal stability and gliding flight.

Hilda was one of many technical experts sent into the wreckage of defeated Germany at the war's end, as there is a report [ix] about her visit to Volkenrode and Göttingen in September-October 1945. These were locations of Germany's equivalent of the Royal Aircraft Establishment, and Göttingen was also of course a university. The Deutsche Forschungsanstalt fur Luftfahrt, at Volkenrode near Braunschweig (Brunswick), was considered to be the most significant developer of high speed flight [x].

Hilda's report is purely technical and does not even mention if she went as part of a group, although it seems most likely that her visit was part of "Operation Surgeon", the postwar programme to inspect what was left of German aeronautical research and practice, and to recruit German technical skills so that they would not fall to the Soviet Union. A very young, new RAE staff member who also went was Ralph Denning [xi], who recalled that the visiting scientists were accommodated in huts hidden in woodlands where the Volkenrode [xii] workers lived whilst working on developing the V2 rocketry. The visitors were apparently stunned by much of what had been developed, including swept-wing aircraft, at that time unknown outside Germany. One of his roles was:

> "editing the Germans' – the Germans did a rough translation of their [work] – if they could and then we had to put it in reasonable English. And there were – I [have] got a record of 1,037 documents that were treated that way. These were either monographs or they were the key reports on swept wings and the

like and the various technologies associated with aeroplanes and engines."[xiii]

As Hilda had spent so long working there before the war, she may have been able to do her own translations of the documents which she was given in Germany. She reported that:

> "The main purpose of the visit was to collect information on problems connected with stability and control, particularly theoretical work and results of systematic experiments, suitable for comparison with theory. The general impression gained from discussions and reports is that the theory of stability and response is not as far advanced in Germany as in this country, but that more work has been done on the measurement of derivatives both on models and in flight and on comparisons between theory and experiment."[xiv]

She was impressed with how much the Germans were able to achieve with very little equipment:

> "This report is concerned almost entirely with work done in Brunswick or Göttingen immediately before or during the war. The most creditable, achievement is the work done by Prof, Schlichting and his collaborators in the Brunswick Technical High School with meagre experimental equipment."

Hawker Typhoon

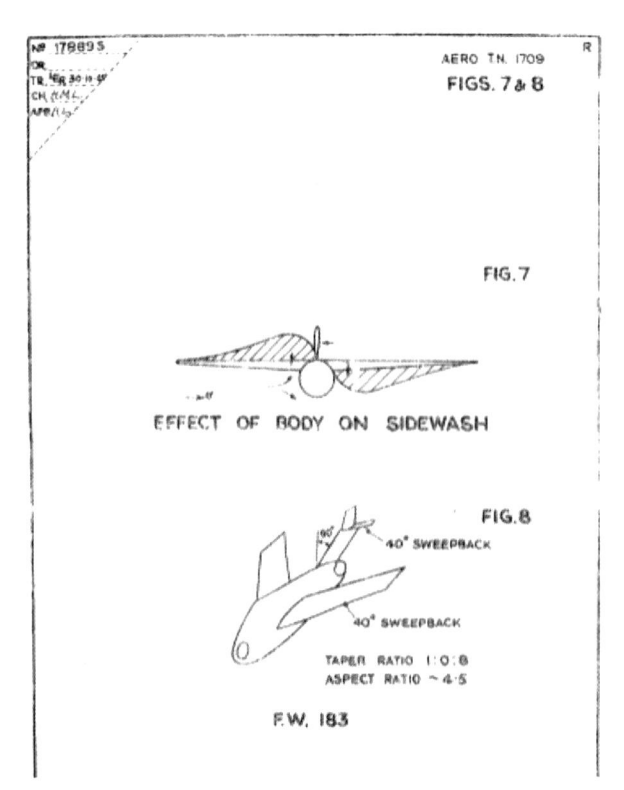

Hilda's sketch from her report of German work on swept-back wings

Her report surprisingly does not mention Professor Prandtl at all, although she seems to have met a number of other researchers: Professors Stupor, Walchner, Ludwieg, Kussner, Jordan, Drescher, Kunze, and Eujen, Dr Siegels and Herren Scherer, Hansen, Kruger, Brennecke at Gottingen and Professors Sohlichting, and Blenk (who led the Institut für Aerodynamik) and Herren Rober and Scholkeneier at Volkenrode. Despite the shorter visit at Volkenrode she still returned with 40 reports, compared with 58 from Göttingen.

Given her own work on the problems of phugoid motion in longitudinal stability, and her paper on the subject in 1942, she would presumably have found Professor Stuper's work on this problem of interest.

Although she reports on the Germans' work on sweptback wing forms, unknown in the UK up to this time, she does not offer any opinion as to the utility of this or any other German work, restricting her report to a factual summary of the work that had been done. However, both the British and the American aeronautical researchers immediately started work on this new form having once been made aware of its possibilities by the people and reports retrieved at this time from Germany.

A year later, one of her last pieces of work published before her death, was an abstract [xv] by Hilda of a report by Hans Trienes on the downwash behind swept-back wings.

Many of the staff from these and other aeronautical research and construction centres in the British Zone of Occupation were recruited to come to Britain to join the RAE, very soon after the war ended and were significant contributors to the UK's postwar aerospace programme.

At the time that Hilda had been in Göttingen in the 1930s, a younger woman – Johanna Weber [xvi] – was taking her mathematics degree there. After the war, Johanna came to the UK with a number of other German aeronautical researchers and spent the rest of her working life at the RAE, eventually helping design the Concorde and Airbus. Johanna's colleague Dietrich Küchemann also went to the RAE and, in 1953 they published 'Aerodynamics of Propulsion' [xvii], based on their work at the AVA, Göttingen from 1940 to 1945 and considered the key text on the topic. The two women may have met when Hilda visited Göttingen in 1945 but they can only have been colleagues very briefly, however, as Johanna arrived in October 1946 [xviii] and Hilda died in the December.

TECHNICAL NOTE AERO. 1709

ROYAL AIRCRAFT ESTABLISHMENT
Farnborough Hants

REPORT ON A VISIT TO VÖLKENRODE AND GÖTTINGEN
SEPTEMBER 28 - OCTOBER 15, 1945

BY

H.M. LYON

DEATH AND LEGACY

Hilda Lyon's grave in Market Weighton cemetery

Hilda Lyon died on the 2nd December 1946, following an operation for an ovarian cyst and intestinal obstructions, and her death was keenly-felt in the aeronautical world as a great loss to the field. Hilda's brother received many heart-felt letters of condolence from her colleagues and friends. Her colleague, W. G. Perrin, noted how she was respected for her painstakingly accurate work [xix]. She had published about 30 papers and official reports, including two on elastic flexing of wings which were published posthumously in 1949. She is buried in her home town, in Market Weighton Cemetery.

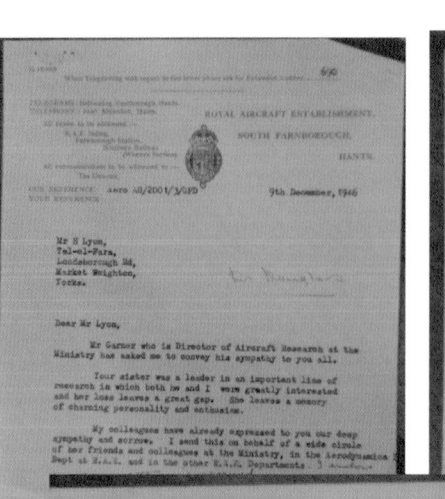

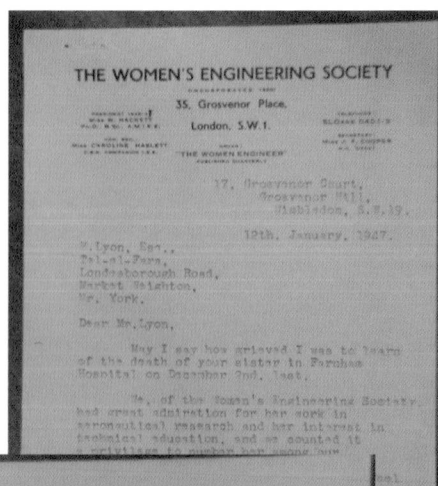

ROYAL AIRCRAFT ESTABLISHMENT,
FARNBOROUGH,
HANTS.

DRAE/27.   5th December, 1946.

Dear Mr. Lyon,

I was very sorry indeed to hear of the death of your sister, Miss H. M. Lyon, and on behalf of myself and the Establishment, I should like to offer sincere condolences in your bereavement.

I, personally, have been associated with Miss Lyon and her work for a good many years, both while she was a member of the Aerodynamics Department, R.A.E., and in her earlier work before joining the Establishment. She won a real place in our affections, and by her painstaking and steady effort she claimed the respect of everyone who was privileged to work with her.

She will be missed by all her friends both at R.A.E. and in industry, and her death is a loss that will be felt throughout the whole scientific world.

Her interests went beyond that of her work, and she took an active part in the education and training of the junior staff within the Establishment, and in this field she made a very real contribution.

Once more – with sincere sympathy,

I remain,

Yours sincerely,

H. W. Lyon, Esq.,
Tel-El-Ford,
Market Leighton,
Yorkshire.

Although she only returned to Yorkshire to live at home at the time when her mother was ailing, she obviously also visited from time to time to spend time with her brother, his family and her cousins in the area. Her great niece, Wendy, daughter of Enid M. Greenwood (nee Lyon), a family member instrumental in documenting Hilda's family genealogy, remembers her as being wonderful fun with her and other children in the family on her visits.

After her death her research, and the "Lyon Shape" which she devised, were incorporated into a class of American submarine USS Albacore, which had the prototype streamlined hull form for almost all subsequent US submarines, and those of many other nations.

The Lyon Shape may even have been in use in the 1944 German midget submarines (Delfins), which suggests that her brief work at Gottingen may have influenced scientists there. The "Lyon shape" term is widely used in the USA but less known elsewhere.

Her 1942 paper, "A theoretical analysis of longitudinal dynamic stability in gliding flight" is considered seminal and continues to be cited in the science of streamlining and boundary layers, although she was of course neither the first nor the last to look at this critical aspect of aeroplane safety. The paper analysed the dynamic and static stabilities of aircraft when undergoing "phugoid" motion. This is when the aircraft pitches up and climbs, and then pitches down and descends, accompanied by speeding up and slowing down as it goes "downhill" and "uphill" (p47), and can be the cause of dangerous instability, especially if any of the plane's control surfaces, such as rudder or ailerons are damaged. Hilda's paper offered various mechanical solutions to dampen this up and down motion and is a key reference in many subsequent papers on phugoid motion problems.

USS Albacore: a 'Lyon shape' submarine

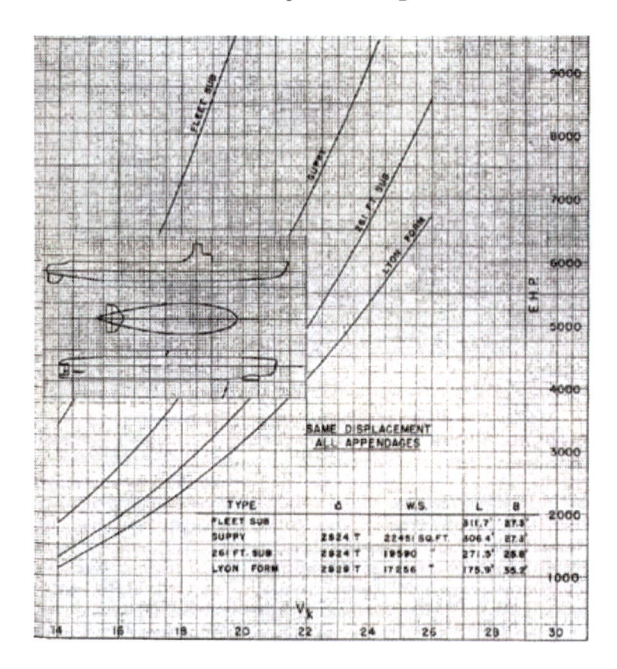

US Navy report on engine power requirements for various submarine hull-shapes, including the Lyon shape

FIG. 15. Effect of a Weight Moment on Phugoid Damping. ($C_L = 1 \cdot 2$).

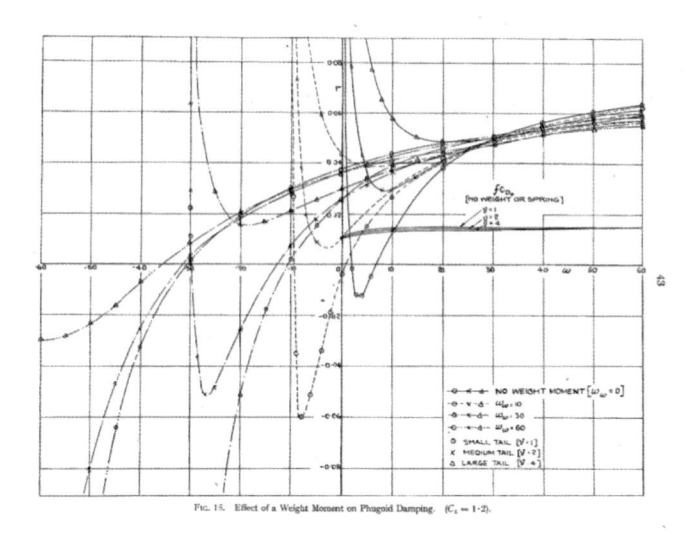

Figure from Hilda's 1942 paper, showing the effects of weight moment on phugoid damping. Title page below.

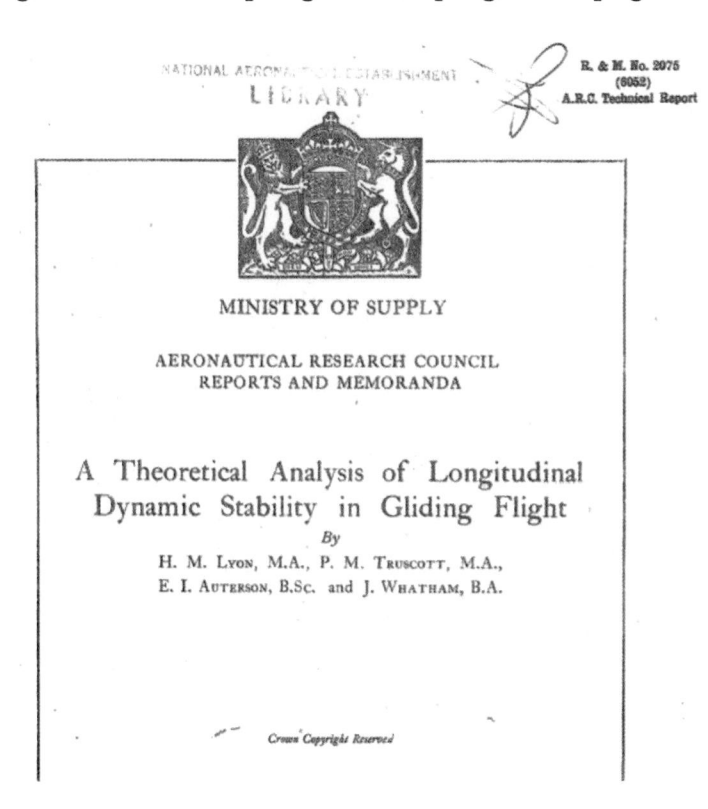

R. & M. No. 2975
(6052)
A.R.C. Technical Report

MINISTRY OF SUPPLY

AERONAUTICAL RESEARCH COUNCIL
REPORTS AND MEMORANDA

A Theoretical Analysis of Longitudinal Dynamic Stability in Gliding Flight

By

H. M. Lyon, M.A., P. M. Truscott, M.A.,
E. I. Auterson, B.Sc. and J. Whatham, B.A.

Crown Copyright Reserved

Today's planes have "anti-phugoid" control software based in part on her work, although it is not always helpful, as in a modern air incident in which knowledge of this phugoid motion was implicated. The US Airways Flight 1549 ditched in the Hudson River on January 15, 2009. Its captain, Chesley ("Sully") Sullenberger, said that the landing could have been less violent had the anti-phugoid software installed on the plane not prevented him from manually getting maximum lift during the four seconds before water impact. Other citations which make use of this last paper of Hilda's work include those in papers on manoeuvrability, drag characteristics of jets and even biomimetics.

A further legacy of her life was a Hilda Lyon Prize for RAE apprentices which was awarded for some years after her death. In her home town a plaque has been erected on the shop where she was born and a room in the newly refurbished town hall is also named in her memory.

The girl from Market Weighton traveled far but her professional legacy traveled much further and she is now becoming better recognised and becoming a great historical role model for potential young engineers today. There is also an entry for Hilda Lyon in the Oxford Dictionary of National Biography, provided in 2019 by Dr Sally Horrocks. Her former school in Beverley has made much of their illustrious alumna and it is to be hoped that this book further expands the circle of those who know of Hilda Lyon's great adventure in aeronautical design.

In 2019 family, friends, aviation historians and local residents attended the unveiling of a plaque to commemorate Hilda Lyon's work and her connection with her home town of Market Weighton. The plaque is on the wall of the general store in Market Weighton's High Street. The ceremony was organised by the town's mayor, Peter Hemmerman, and the plaque officially unveiled by the Honourable Dame Elizabeth Susan Cunliffe-Lister, DCVO, the Lord Lieutenant of East Yorkshire. Many people attended, including teachers and pupils from Beverley High School and nine members of the Lyon family who had traveled from the USA and Australia.

REFERENCES

i. Greenwood, E. M 2000. Market Weighton. Changing Face and Faces. Highgate Publications (Beverley) Lt. ISBN 1-902645-13-8. This book was written by Hilda's niece, the late Enid Greenwood and contains a detailed history of the family and its grocery firm.

ii. Private communication from the archivist at Beverley High School.

iii. Hilda Lyon's Roll card, kindly provided by Newnham College archives.

iv. Information about her early career comes from her application form to become an Associate Fellow of the Royal Aeronautical Society in 1922, kindly provided by their archives.

v. Lyon, H. 1944. Adventures in Aeronautical Design. The Woman Engineer, volume 5, issue 19. Available from https://www.theiet.org/resources/library/archives/exhibition/women/wes-journal.cfm? Apart from her official reports, this talk to members of the Women's Engineering Society, as reported in its journal, seems to be the only publication by Lyon about her own life and work. All quotes from Lyon in this paper come from that article.

vi. This may have been the occasion of the photograph of staff 'playing the roles of passengers at dinner' in C.P. Hall's article "R101 passenger accommodation, part" in Dirigible Volume 64, 2011, p18. None of the photos are clear enough to be sure who the various women might be.

vii. The effect of turbulence on the drag of airship models by Hilda M Lyon. Thesis. Massachusetts Institute of Technology, 1932. Also: Google Scholar citation check.

viii. Ludwig Prandtl's publications list. Accessed 13 June 2020: http://www-history.mcs.st-and.ac.uk/Extras/Prandtl_publications.html

ix. Lyon H M. Oct 1945. Aero/TN/14 Report on a visit to Volkenrode and Gottingen. Sept 28 - Oct 15, 1945. 1709 ARC 9348. Copy held at FAST archives

x. Nahum, A. 2003. "I believe the Americans have not yet taken them all!": The Exploitation of German Aeronautical Science in Postwar Britain. Conference proceedings: Tackling Transport, 2003, pp99-138.

London Science Museum, p102.

xi. Ralph Denning: a visit to post-war Germany as part of T-Force. Voices of Science. British Library. He was born in 1925, so was only 20 when sent to Germany.

xii. For a history of Volkenrode, pictures of the facility then and now and images of the wind-tunnels, see Accessed 16th November 2018.

xii. Transcription Ralph Denning Page 63 C1379/68 Track 4. Accessed 16th November 2018 from

xiv. Aero/TN/14, page 2

xv. Aero/TN/1819a. October 1946.

xvi. Weber was later a significant contributor to the Concorde project.   Accessed 16th November 2018.

xvii. Küchemann, Dietrich; Weber, Johanna (1953). Aerodynamics of Propulsion. McGraw-Hill publications in aeronautical science (1 ed.). New York, NY: McGraw Hill.

xviii. Nahum, A. 2003. "I believe the Americans have not yet taken them all!": The Exploitation of German Aeronautical Science in Postwar Britain. Conference proceedings: Tackling Transport, 2003, pp99-138. London Science Museum, p114.

xix. Deacon, K. The Lyon Shape. Langrick Publications. ISBN 0-9456606-3-3. This compilation of cuttings and photos, largely from the family's archives, maybe available via the Howden Heritage Centre but the museum itself is now closed.

xx. Bradfield, F. Newnham College Roll, Jan 1947.

HILDA M. LYON'S PUBLICATIONS, IN CHRONOLOGICAL ORDER

1930 The Strength of Transverse Frames of Rigid Airships, Hilda M. Lyon. 1930. The Aeronautical Journal, Volume 34, Issue 234. June 1930 , pp. 497-5561

1932 The effect of turbulence on the drag of airship models by Hilda M Lyon. Thesis. Massachusetts Institute of Technology. There have been 20 citations from 1934-2014. dspace.mit.edu/bitstream/handle/1721.1/58438/36135138-MIT.pdf;sequence=2

1933 Effect of turbulence on drag of airship models, by Hilda M. Lyon. London, E.N. Stat. Off., 1933. 28 p . diagrs., tables. (A.R.C. R. & M. no. 1511)

1934 The Drag of streamline bodies. The relative importance of skin friction and pressure in relation to full-scale design, by Hilda M. Lyon. Aircraft engineering, London, Sep. 1934, v. 6, no. 67, p. 233-39.

1935 A study of the flow in the boundary layer of streamline bodies. HM Lyon - 1935 - HM Stationery Office. There have been 8 citations from 1939-2012

1936 On the force between two coaxial single layer helices carrying current. Sir Richard Glazebrook, F. R. S., H. M. Lyon, M. A. Proc RoySocA Maths, Physics and engineering sciences 2 March 1936.DOI: 10.1098/rspa.1936.0032. This is a summary - Full paper was printed in the Collected Researches of the National Physical Laboratory, vol. 24 . Although Glazebrook is given as first author he credits her: "My share amounts to little more than the evaluation of the integral expression for the force and suggestions as to the method of treatment; the integration is due to Miss Lyon.—R. T. G."

1936 Oscillations of elastic blades and wings in an airstream. Duncan, William J.; Collar, Arthur R. ; Lyon, Hilda M. [London] : [HMSO], 1936. There have been 6 citations from 1937-1949.

1937 Calculated flexural-torsional flutter: characteristics of some typical cantilever wings. Duncan, William J.; Lyon, Hilda M. [London] : [HMSO], 1937. There have been 15 citations from 1943-1983.

1937 A review of theoretical investigations of the

aerodynamical forces on a wing in non-uniform motion. Lyon, Hilda M.[London] : [HMSO], 1937. There have been 8 citations from 1943 – 2013.

1938   Torsional oscillation of a cantilever when the stiffness is of composit origin. Duncan, William J.; Lyon, Hilda M. [London] : [HMSO], [1938]

1938   The Effect of Ground Interference on the Trim of a Low Wing Monoplane. HM Lyon, JE Adamson - 1938 - HM Stationery Office

1938   Credited by HB Squire in The lift and drag of a rectangular wing spanning a free jet of circular section The London, Edinburgh, and Dublin Philosophical Magazine and Journal of Science. Series 7, Volume 27, 1939 - Issue 181, Pages 229-239 [" I am indebted to Professor L. Zosenhead and to Miss HM Lyon for advice and criticism."]

1939   Wind-tunnel tests on high-lift devices by H.M Lyon (Book) 1 edition published in 1939

1939   Lift Increase by Boundary-Layer Control. HM Lyon, R Hills – RAE BA Dept. Note WT, 1939

1939   Aerodynamical derivatives of flexural-torsional flutter of a wing of finite span. HM Lyon, WP Jones, SW Skan - 1939 - HM Stationery Office

1941   Lateral Instability and Rudder-Fuselage Flutter. Rep. No. AD 3164. HM Lyon - British RAE, May, 1941

1942   A theoretical analysis of longitudinal dynamic stability in gliding flight by Hilda M Lyon (Book). Authors Hilda M. Lyon, P. M. Truscott, E. I. Auterson, J. Whatham. Publisher H.M. Stationery Office, 1942.

1944   Volume 2027 of Reports and memoranda, Great Britain Aeronautical Research Council. Authors Sidney B. Gates, Hilda M. Lyon. Publisher   H.M.
Stationery Office, 1944.

1945   Aerodynamic forces on wings in simple harmonic motion. WP Jones - 1945 - naca.central.cranfield.ac.uk. This includes Lyon's Theory.

1949   A method of estimating the effect of aero-elastic distortion of a swept-back wing on stability and control derivatives by Hilda M Lyon. There have been 8 citations from 1951-1982.

1949   Aerodynamical derivatives of flexural-torsional

flutter of a wing of finite span. Calculations of derivatives for a wing of finite span by H. M Lyon

Technical Reports for ARC

1944   Aero/RTN/39   A continuation of longitudinal stability and control analysis. Gates S B, Lyon H M. Feb 1944. 1912. ARC 7687   R&M 2027

1944   Aero/RTN/41A. A continuation of longitudinal stability and control analysis Part 2: Interpretation of flight tests. Gates S B, Lyon H M. Aug 1944. 1953

1945   Aero/RTN/36   Addendum to BA 1755 A theoretical analysis of longitudinal dynamic stability in gliding flight. Lyon H M. Jan 1945. 1755. ARC 8374

1945   Aero/RTN/43A A general survey of the effects of flexibility of the fuselage, tail unit and control systems on longitudinal stability and control. Lyon H M, Ripley J. Jul 1945. 2065. ARC 9015   R&M 3415

1945   Aero/TN/14 Report on a visit to Volkenrode and Gottengen. Sept 28 - Oct 15, 1945. Lyon H M. Oct 1945. 1709. ARC 9348

1946   Aero/TN/16 The estimation of aerodynamic loads on swept back wings. A summary of the present position (May 1946). Lyon H M, Gdalihu M . Jun 1946. 1. 797.   ARC 9951

1946   Aero/TN/17 A method of estimating the effect of aero-elastic distortion of a swept back wing on stability and control derivations. Lyon HM. Jul 1946. 1813. ARC 10024   R&M 2331

1946   Aero/TN/17 Systematic measurements of downwash behind swept back wings by Hand Trienes A.I.T.B. Report 45/8 Abstract.   Lyon H M.   Aug 1946. 1819   ARC 9959

1946   Aero/TN/17 Addendum to R.A.E. Technical Note Aero 1719. Lyon H M. Oct 1946. 1819(a). ARC 10062

[The title of this book has been taken from that of her own talk as transcribed below.]

ADVENTURES IN AERONAUTICAL DESIGN AND RESEARCH

Informal talk to the Women's Engineering Society, 4th March 1944, by Hilda M Lyon, as published in The Woman Engineer

I hope that my title will not lead you to expect thrilling stories of flying experiments or of outstanding scientific discoveries. Lots of the work is very humdrum, just pushing a slide rule or reading wind tunnel balances. I can assure you that aeronautical design and research has its full share of patient slogging donkey work. There are times when a week's or a month's work may go into the waste paper basket because the basic design has been altered. The work of a year and more may be put away on a shelf because someone in authority decides that that particular design is not wanted after all, or that a research investigation is not of sufficient urgency or promise at the moment to warrant more time or money being spent on it. On the other hand there are exciting times when the prototype flies or when a long piece of research gives us an answer to some perplexing problem or points the way to new developments. The job is never finished.

The more successful research may be the more it leads to new ideas and possibilities and one is learning, learning all the time. Of nothing is it truer to say that the more one knows, the less one thinks one knows.

The first adventure

It was the last war that gave me an entry into aeronautics as stress merchant or technician in an aircraft firm. The application of mathematics to engineering problems appealed to me much more than the study of pure mathematics, but physics and chemistry were left out of my education and I entered the engineering world with only an academic training in mathematics. For a shy country bumpkin like myself that was the greatest adventure of all. It was like going to a foreign country where men talked a different language, the language of the practical engineer instead of that of the mathematician. If I could start at the beginning again I should go to the local technical school with the works apprentices, failing anything better, but shyness and I'm afraid a sense of superiority after three years at Cambridge, prevented me. Also of course I had to read up a lot of structural and aerodynamic theory to enable me to do my particular job.

For those who are familiar with modern stress offices I should like to explain that during the most of 5 years in 2 firms I was the stress office and the aerodynamics department, responsible only to the chief designer for my work, and with no assistant to check it. If an experienced draughtsman thought that a flange ought to be 1/4" thick and I said it ought to be 1/2" I went back to check my calculations. When I was right I had the job of explaining in his language why he was wrong before I could convince him. And so gradually I learnt to understand and speak the engineering language, thought doubtless with a strong mathematical accent.

Structures were relatively easy then, of course, mostly braced biplanes with no stressed skin to worry about. My first job was on a biplane bomber of about the same span as the Lancaster, getting ready to bomb Berlin in 1920 or so, but the war was over before even the prototype flew. Later efforts included a tri-plane and what I believe was the first metal monoplane designed in this country, but it never got beyond the drawing board stage. That was my first experience of the scrapping of the work of a year or more, with nothing to show for it but the experience gained.

It was about the same time that I learned something about propeller design and stressing. I remember that we designed 2 propellers for a light aeroplane competition, one for high speed and the other of a slightly smaller pitch for take-off. That was, of course, before the days of variable pitch propellers and when the difference between top speed and take-off speed was small compared with modern standards. The aeroplane won the speed race with the propeller designed for take-off. That destroyed my faith in my calculations of blade settings but later I learnt that more expert propeller designers had had similar experiences.

The work interested me but there seemed to be no prospect of more promotion and responsibility for a woman mathematician. So in a light-hearted moment my sister and I threw up our respective jobs and went to Switzerland for 6 weeks, then returned with emptier pockets to look for new ones.

Airships

My next job was on airship stressing in the technical office at Cardington when the R101 was being designed for carrying passengers to the Dominions and India. Airship structures are very different from braced biplanes and much more interesting and complicated. They have more in common with modern aeroplane structures. Airships had a greater fascination for me than aeroplanes have ever had. There was the interest of watching the ship growing in the hangar and later on the excitement of the early flights, particularly of my one joy ride as a passenger. After a week of waiting for suitable weather, R101 was pulled out of the shed for the first time one morning, before sunrise, for the first flight. She was moored to the mast over the weekend, gleaming in the sunshine of a fine October Sunday. All the roads for miles around were blocked with cars of sightseers from all parts of the country and the quickest way to get from Bedford to Cardington was to walk. A month later she was riding out a fierce November gale with gusts of up to 80mph and was swung round the mast by a sudden change of wind direction of $135°$ in a line squall.

The R100 was built at the same time in Yorkshire and came to Cardington after her first flight. She was towed into the shed after dark by the light of searchlights; with the strong lights and shadows the ship looked twice the normal size. R100 later flew to Canada and back and I watched the return to the mast on my last day at Cardington before sailing for America for my next adventure. This too was concerned with airships.

In the early days of their design, R101 and R100 were the last word in streamlining. By the time they were flying about 4 or 5 years later we were beginning to realise that the tunnel tests on small models were misleading and that the thin streamlined nose and tail were unnecessary: a somewhat bluffer shape would give as low a drag on the full scale airship, while providing more valuable space for lifting gas to support the heavy mooring gear and fin structure in nose and tail. Such an arrangement would have reduced considerably the bending moments in the whole ship and the weight of the structure.

Anyone engaged in aeroplane stressing knows something of the importance of keeping the weight down to the minimum without sacrificing the necessary strength. It is much more important on airships. I was particularly interested in this problem and decided to combine a trip to America with an attempt to solve it by doing wind tunnel tests on airship models at the Massachusetts Institute of Technology. This may sound crazy, as I have just told you that the wind tunnel tests were misleading but by this time I knew what I was looking for and I deliberately faked the conditions in the tunnel to give results which could be connected with more confidence to full scale conditions. I eventually stayed a second year on the research staff at MIT to complete my experiments and am very grateful for the opportunity and encouragement given me by the staff of the Aeronautics Laboratory.

I had been only a month in the USA when I heard of the fatal accident to the R101 on an attempted flight to India. That was a hard blow and it took all the heart out of my research for a time.

Airships were very much in the news then. During my early days at Cardington a small Italian-built airship, the Norge, flew across the North Pole and landed safely in Alaska. It called at Pulham in Norfolk for fuel and gas on the way from Italy to Oslo with the designer, Nobile, and an Italian crew and some Norwegians on board. A party of us went to see the ship and waited to see them leave about midnight.

A later visitor to Cardington was the Graf Zeppelin, probably the most successful airship that has ever flown. It had a range of about 6000-7000 miles with 20 passengers, a crew of 40 and about 15 tons of freight. It had several long flights to its credit including a flight round the world.

While I was in America two large airships were being

built for the US Navy, rather larger than R101 and R100. I saw the Akron during its construction and again in the air. It, too, crashed about two years later and also the Macon, its sister ship, but that was some time after my return to England. Another interesting American airship which was flying while I was there was a small all-metal ship, popularly known as the tin ship. There were plans at the time to build larger ones with metal skins.

In spite of all these disappointments, followed later on by the loss of the newest German airship, the Von Hindenburg, I do not regret that I played a part in airship development.

In the USA

America gave me my first taste of experimental work and of controlling electric motors and using a lathe in the machine shops. After a preliminary course in wind tunnel testing, I was allowed to use the tunnel for my own research work, sometimes with assistance but often alone, particularly in the evenings when the rest of the staff had gone home. It was a bit eerie sometimes. There were some sheets of duralumin lying in a corner and these used to creak and move about in the return flow from the tunnel when I was running it at full speed. Modern tunnels running at higher speed and with more elaborate apparatus cannot, of course, be left to the tender mercies of a novice such as I was then, but the experience in that tunnel gave me more confidence and taught me more than assisting an expert staff in a more up-to-date tunnel could have done. Fortunately I did not have an accident.

We have nothing in this country which quite corresponds with MIT. It is both a research institute and a technical college of university standing, giving its own degrees which rank among the highest in the States in science and engineering. The Guggenheim Aeronautical Laboratory is the home of the aeronautical engineering department. It gives courses in aircraft design, structures, aerodynamics, wind tunneling etc; each year a band of students would build their own glider and take it down to Cape Cod to fly, break it up, bring it back, patch it up and try again.

Of my aeronautical visits the most interesting were trips to Lakehurst, New Jersey and Akron, Ohio, to see existing and future airships and to Washington and Langley Field, the research station of the National Advisory Committee for Aeronautics. Washington to me will always mean lovely spring weather after a long New England winter and cherry trees in bloom on the banks of the Potomac. To Langley Field one travels overnight by steamer from Baltimore down Chesapeake Bay to the coast of Virginia. That was the only place I visited where negroes and white people sat in different compartments of the tramcar. In Boston one's neighbour might be of any colour and nationality and speak almost any language.

Germany's Fateful Year

I followed up my two years in America with a year at Göttingen in Germany. In aeronautical circles, Göttingen is famous for its research institute, Kaiser Wilhelm Gesellschaft fur Strömingsforschung and more so for Professor Prandtl, the head of the institute and a professor of Göttingen University. His own research work covers a wide field and he has taught and inspired countless students from all over the world. The total contribution of his school to the science of aeronautics is a very large one. My own work was mainly study and theoretical research with advice and help from Professor Prandtl and others of his staff.

The political developments during my stay in Germany in 1932-33 are of more general interest to most of you. When I first went, Germany was having a new government every two months or so and none could get a majority or do anything constructive. The only two parties who knew what they wanted were the National Socialists or Nazis and the Communists; as the lesser of two evils the most influential of the rest chose the Nazis, and Hitler was made Chancellor in January 1933. A large proportion of the students were Nazis and they were jubilant. The Fackelzug or torchlight procession in honour of Hitler was the largest the Göttingen had known for years, if ever. It was picturesque and impressive to an onlooker like myself, but there was something ominous about it as these serious looking lads marched past singing the Horst Wessel song. Later on I was in Nuremberg for a Nazi celebration but I never saw or heard Hitler except on the radio.

It was not until after the Easter vacation that the effect of the change was marked in university circles. During the vacation the Göttingen synagogue was raided, and there was the usual burning of books quite early on.

The summer session of the university began a fortnight late, all the Jewish professors having been beurlaubt, sent on leave. They were paid their salaries for a time, I believe, but were not allowed to lecture. This eliminated most of the mathematical and physics faculties, including very distinguished men and Nobel prize winners. Even the Nazi students were distressed when their own professor was dismissed in effect and there was in general a feeling of respect and affection for the Jews they knew personally. But at the same time they felt that Jews as a race had got all the best professional jobs and left none for the non-Jews. By the time I left in July the opponents of Hitler, non-Jewish ones, were learning to keep their mouths shut and to accept the Hitler regime at least outwardly.

Interlude

My next four years were spent at home for family reasons, nursing and housekeeping and carrying on with my work at the same time. Some of it was paid, much of it was not, but the main thing to me was to keep in touch with aeronautical work and to make what contribution I could. The university libraries of Hull (20 miles away) and Leeds (40 miles) were a great help and rare visits to the National Physical Laboratory in Teddington and RAE at Farnborough went some way to keep me in touch with other work. But I badly missed the constant contacts and discussions with colleagues.

Back to Farnborough

I was very glad to get back to full time aeronautical work again in 1937 when I came to the Aerodynamics Department at the RAE Farnborough. At first I was doing experimental work in the wind tunnels but I feel more at home on my present more theoretical work. I enjoyed the practical work and am glad to have had the experience, but I have no wish to return to it now.

The rest of the story will have to be told after the war if it is worth telling at all.

IMAGE CREDITS

Most images are from the public domain, wikicommons or similar, or are author's own. Others as follows:

Family archive of the Lyon family kindly shared by Wendy Smith.

Cardington airship shed: This file is licensed under the Creative Commons Attribution 2.0 Generic license.

Newnham College Cambridge. A. D. White Architectural Photographs, Cornell University Library Accession Number: 15/5/3090.01108 The digital file is owned by the Cornell University Library which is making it freely available with the request that, when possible, the Library be credited as its source.

The Woman Engineer, journal of the Women's Engineering Society, the first 18 volumes of which are digitised and freely available online:

THE AUTHOR

Nina Baker has had a varied career, having become a merchant navy deck officer on leaving school and taken an engineering design degree in her 30s, from the University of Warwick. She then gained a PhD in concrete durability from the University of Liverpool. She has lived with her family in Glasgow since 1989, working variously as a materials lecturer in further education and as a research administrator and, until 2017, as an elected city councilor.

Now retired from all that, her interest in promoting STEM careers for girls has led her to become an independent researcher, mainly specialising in the history of women in engineering, bringing forgotten stories out into the light.

This book is the result of discovering Hilda Lyon's amazing story quite a few years ago and then having the chance to visit her home town and meet some of her relatives. Market Weighton has now put up a plaque in the High Street to commemorate Hilda, its brilliant daughter.

If you have any new pieces of information relevant to Hilda Lyon's life, Nina Baker can be contacted by way of her blog:
https://womenengineerssite.wordpress.com/

Printed in Great Britain
by Amazon